The Periodic Table of the Fourth Industrial Revolution

147 Terms and Concepts:
Organized and Explained

Stephen Haag

Janus Press, LLC

Copyright © 2022 Stephen Haag and Janus Press, LLC

All rights reserved, including the right of reproduction in whole or in part in any form.

The Periodic Table of the Fourth Industrial Revolution is protected by USPTO trademark serial number 97412226, applied for May 16, 2022.

Janus Press, LLC (www.januspressllc.com)

Printed in the United States of America
First Printing: July 2022

Library of Congress
ISBN: 979-8-9857439-1-3

Periodic table images provided by Anton Prikhodko at www.upwork.com.

Cover design by Mikola Shelepa at www.upwork.com.

Janus Press books are available at a special discount for bulk purchases. For more details, email info@januspressllc.com or contact the author directly at techbookwriter@gmail.com.

DEDICATION

For my Mom and Dad, Iona and Carl.

The world is indeed a better place because of your love.

CONTENTS

	Introduction	1
1	Energy Harvesting & Storage Technologies	5
2	Communications Technologies	15
3	Sensing Technologies	21
4	Technology Architectures	31
5	The Internet of Things	37
6	Artificial Intelligence	41
7	Blockchain	49
8	Cryptocurrency	55
9	3D Printing	59
10	Extended Reality	63
11	Super Technologies	67
	Index	77

INTRODUCTION

The periodic table of the fourth industrial revolution is a tabular display of the key elements, concepts, technologies, and terminologies associated with and that are vitally important to the success of the fourth industrial revolution.

In case you missed the last 275 years, please allow me to summarize.
- 1st Industrial Revolution (1750s-ish until the 1840s-ish): The use of water and steam to power mechanized factories. This led to unprecedented increases in worker productivity over non-mechanized processes, often as much as fifty-fold for daily output. Other advancements included chemical manufacturing, iron production, the steam engine locomotive, cement, glass making, and gas lighting.
- 2nd Industrial Revolution (19870s-ish until around 1914): The use of electricity, power grids, the internal combustion engine, the telephone, and the telegraph. This created a shift from water and steam power to electricity and the gas-powered internal combustion engine. Iron and steel production processes became much better. Paper-making processes greatly improved, leading to an increase in general education

(reading and writing). The use of ammonia in fertilizer led to unbelievable crop yields.
- 3rd Industrial Revolution (1950s-ish until today): Punctuated by computing and digital technologies, communications technologies, smartphones, and so on. This era has been referred to by many names… the digital revolution, the digital age, and the information age.

Today, we are entering the fourth industrial revolution, highlighted by new technological advancements and their uses.

The periodic table of the fourth industrial revolution includes 11 major blocks:
1. Energy Harvesting and Storage Technologies
2. Communications Technologies
3. Sensing Technologies
4. Technology Architectures
5. The Internet of Things
6. Artificial Intelligence
7. Blockchain
8. Cryptocurrency
9. 3D Printing
10. Extended Reality
11. Super Technologies

I would make some qualifying remarks before we dive in.

WARNING: We're not really going to "dive in." If you want an in-depth, technically perfect review of all these technologies, this book is not for you. But, if you're looking for a nice overview of the terms and concepts associated with the fourth industrial revolution, this book will do the job.

Secondly, some of the terms in the periodic table are not unique to the fourth industrial revolution. WiFi, for example, has been around for many years. But, its use and continued development is essential for the success of many fourth industrial revolution technologies.

Thirdly, if you're familiar with the periodic table of elements, you know that each element is unique, an individual element dependent on its atomic number and various properties. Not so for the periodic table of the fourth industrial revolution. While many are individual elements such as a sound sensor, others are collections of elements, such as an autonomous vehicle. The periodic table of the fourth industrial revolution is a collection of terms, some elemental in nature and some only possible as a collection of those other elements.

Fourthly, not all terms are strictly "technological" in nature. The Paris Agreement, for example, isn't a technology but rather our collective attempt to address climate change. I've included The Paris Agreement and many other non-technology terms as they are and will continue to be important as we witness the unfolding of the fourth industrial revolution.

Fifthly, it has been a real challenge to explain each term in just a few sentences. Indeed, many of the terms such as artificial intelligence (AI), have no generally agreed-upon definition. If you have a better way to explain a particular term, please do contact me. I'd love to hear from you.

And finally, this is a black-and-white book. So, please refer to the front cover for the complete (and colorful) periodic table of the fourth industrial revolution. The back cover contains the legend of major blocks.

The Periodic Table of the Fourth Industrial Revolution

Energy Harvesting & Storage Technologies

3DP						ST	AV	DAT	LC	D						XR	AR	VR
IoT	FDM	F		3DB		AD	eVTOL	W3O	Mv	Ro						VRH	MR	MRH
IoCT	DLP	R	3DMS	STLF	SIS	C	PKC	PK	PYK	CE	CW	SiC	NFT			Hg	A	SM
SC	IoET	SH	IoAT	GC	BA	3DR	Bc	BN	BHK	M	I	Dapp	SCt			DeFi	Fi	DAO
AI	SG	IIoT	LMAI	AGI	TMAI	SAAI	ASI	ANN	ML	SL	UL	DL	GA			Ne	Bmm	NLP
	ANI	RAI																

						TA	Q	QC	N	Nt	DLT	CC	EC	FC	DT			QRC
CT						RFID	NFC	EMV	WiFi	LiFi	5G	6G	Lt	Tt	HFT		DST	UPC
SsT	BT	HT	SS	CIC	SR	SS	CV	S	L	Rdr	36OC	OCR	SLAM	B			ET	PA
E	GhG		CIC	CZ	CN		RE	PVS	MoSo	WF	GP	HFC	Pe				MC	EVB
BTU	GE		GW				SP	CSP	WP	Hp	Bm	OP	Ba				PB Brc	AB GR	LIB
							EH										SB	SSB

4

CHAPTER 1
ENERGY HARVESTING AND STORAGE TECHNOLOGIES

Energy [E] – the ability to do or the capacity for doing work or undertaking an activity. When you rub your hands together, you are using kinetic energy, which in turn creates thermal energy. Think about a simple flashlight. The energy inside the battery is stored in chemical form. The chemical energy is converted into electrical energy to power the bulb. The bulb uses the electrical energy and converts it into radiant energy.

In this manner, energy is actually never *consumed* in the traditional sense of consumption. That is, one form of energy simply becomes another form of energy. For example, when you eat, the food contains chemical energy which your body stores and eventually converts into kinetic energy when you move. Likewise, when you drop something, the gravitational energy that pulls it to the ground is converted into kinetic energy.

Energy can take on many forms including:
- Heat/Thermal
- Light/Radiant
- Motion/Kinetic
- Electrical
- Chemical

- Gravitational
- Magnetic
- Mechanical (many forms including pressure)
- Elastic… just to name a few.

British thermal unit [BTU] – the amount of heat required to raise the temperature of one pound of water by one degree Fahrenheit. (One pound of water is just shy of 16 ounces, or 2 cups of water, or really close to a half liter.) BTU, as a term, has been around since the 1800s, so it's by no means specific to the fourth industrial revolution. However, it is the common measure to which we convert the measures of all other energies. For example, 1 BTU is 0.293 watts. In 2020, the United States consumed 11.6 quadrillion BTUs of all types of energy (petroleum, coal, nuclear, natural gas, and renewables). [See https://www.eia.gov/energyexplained/us-energy-facts/]

The focus of the fourth industrial revolution is on renewable energy.

Renewable energy [RE] – energy that comes from naturally-replenishing and non-finite resources. Given enough time, we would eventually exhaust the earth's supply of coal, oil, and other fossil fuels. That is to say, they are not a form of renewable energy. However, renewable energies are (mostly) inexhaustible. Just think about wind power. If you capture wind power today, that does not diminish the amount of wind in the future. As well, the use of renewable energies can help us address climate change.

Greenhouse gas [GhG] – a gas that emits radiant energy causing the greenhouse effect. The primary greenhouse gases are water vapor, carbon dioxide, methane, nitrous oxide, and ozone. Of those, carbon dioxide is the most prevalent in causing global warming.

Greenhouse gases, such as carbon dioxide, are actually good; we don't want to get rid of them all. In fact, if we had no greenhouse gases, the earth's temperature would plummet to about 0^0 F (severe and catastrophic global cooling).

Greenhouse effect [GE] – the warming of the earth's surface from radiation in the earth's atmosphere.

Climate change [ClC] – long-term shifts in temperatures and weather patterns. Climate change includes both global warming and global cooling.

Global warming [GW] – the increase in the overall temperature of the earth's atmosphere, mainly occurring because of the greenhouse effect through the increase in carbon dioxide levels. To address the increase in carbon dioxide levels, much effort is underway in carbon zero and carbon neutral technologies and processes.

Carbon zero [CZ] – no carbon emissions as a byproduct of the use of a product or service, which does include the use of energy. Examples include a wind farm generating electricity or a battery deploying electricity.

Carbon neutral [CN] – removing as much carbon dioxide from the atmosphere as is produced from the use of a product, service, or energy. This creates a carbon zero state but does require carbon sequestration to remove as much carbon dioxide as is being emitted.

Carbon sequestration [CS] – removing carbon dioxide from the earth's atmosphere. There are many approaches such as filtering the air, removing the carbon dioxide, storing the carbon dioxide in tubes, and burying those tubes in the ground. Carbon sequestration techniques are required when the focus is being carbon neutral.

Energy harvesting [EH] – the capturing of a renewable energy at its source (e.g., the kinetic energy of wind or the kinetic or gravitational energy of water movement) and converting it into another energy form. The latter energy form is most often electricity but can also be heat. Energy harvesting is in contrast to energy manufacturing, the process of extracting energy from oil, coal, and other fossil fuels.

The major types of renewable energy include:
1. Solar power
2. Wind power
3. Hydropower (water power, hydroelectric)
4. Geothermal power
5. Biomass
6. Hydrogen fuel cell
7. Ocean power
8. Piezoelectricity

Solar power [SP] – the conversion of sunlight (radiant energy) into electricity. That's the most well-known application of solar power and includes photovoltaic systems, concentrated solar power systems, and mobile solar units. Solar power can also be in the form of capturing the thermal energy of the sun for various uses including solar water heating, the heating of a building, desalination, and solar water disinfection (SODIS).

Photovoltaic system [PvS] – the use solar panels to absorb and convert sunlight into electricity. Solar panels on a home are a photovoltaic system. Commercial photovoltaic systems can occupy several acres of land.

The Periodic Table of the Fourth Industrial Revolution

Concentrated solar power [CSP] – the use mirrors and lens to concentrate a large area of sunlight onto a receiver, which converts it to heat which drives a heat-engine/steam turbine connected to an electrical power generator. Concentrated solar power units often occupy several acres of land and are circular in form.

Mobile solar [MoSo] – small portable solar power units that are easy to carry and transport. Usually no more than about 1-foot square, MoSo units are effective for charging portable devices such as a smartphone.

Wind power [WP] – the capturing of the kinetic energy of wind to power wind turbines which provide mechanical power to electric generators which in turn generate electricity. Wind power units can be onshore (i.e., on land, usually in rural areas) or offshore (i.e., on water). The combination of several wind turbines in a wind power unit is called a wind farm.

Wind farm [WF] – a collection of wind turbines in close proximity that power numerous electric generators that are usually connected to an electricity power transmission network. The power transmission network moves electricity from the energy source (the wind farm) to homes, businesses, cities, and grid storage units.

Hydropower (water power) [Hp] – the capturing of the kinetic or gravitational energy of running or falling water to run a turbine or series of turbines which in turn generate electricity. This is achieved in a hydroelectric power plant. Hydroelectric power plants are often thought of in conjunction with human-made dams that create a reservoir of water so there is a consistent flow of water to the turbines. But, they can also be run-of-river hydroelectric power plants that rely on a consistent flow of river water to meet energy demands.

The Periodic Table of the Fourth Industrial Revolution

Geothermal power [GP] – can take on two forms, either (1) electricity generated from geothermal energy or (2) heat generated from geothermal energy. In either case, the naturally-occurring thermal energy in the earth's crust is used.

Biomass [Bm] – the use of plant and animal material as fuel to generate electricity or heat. Biomass, specifically wood, was the primary source of energy in the U.S. up through the mid 19th century. Biomass includes:
- Biomass waste – waste from forests, yards, and farms (bark, sawdust, wood scrap, etc.)
- Biofuels – energy crops or crops grown specifically for energy production (willow trees, poplar trees, and elephant grass are examples); ethanol and biodiesel also fall into this category
- Wood – the burning of wood for cooking, lighting, and heating

Hydrogen fuel cell [HFC] – produces electricity by combining hydrogen and oxygen atoms. A hydrogen fuel cell is 2 to 3 times more efficient than a traditional internal combustion engine running on gasoline. The only byproduct is water, so it falls into the category of carbon zero.

Ocean power [OP] – produces electricity by capturing the energy of the ocean. This can occur in two ways. Firstly, heat from the ocean's surface and cold water from ocean depths can be combined to run electric generators. Secondly, electricity can be generated by capturing the kinetic energy (ebb and flow) of the ocean. Both are in their very early stages of research.

Piezoelectricity [Pe] – the capturing of pressure applied to specific types of materials (for example, certain crystals and ceramics) to create electricity. For example, piezoelectric energy harvesting plates can be embedded in a sidewalk. Pedestrian traffic on the sidewalk could be captured and converted into electricity.

Paris Agreement [PA] – (also called Paris Accords, Paris Climate Agreement, or Paris Climate Accords) an international treaty on climate change adopted in 2015 by 197 countries. The Paris Agreement includes:
- A main goal of reducing greenhouse gas emissions in an effort to limit global temperature increases in the 21st century to no more than 2 degrees Celsius above preindustrial levels.
- A commitment by developed nations to assist less developed countries in an effort to address climate change, through financial resources and the sharing of technologies.
- A transparent framework for monitoring and reporting the efforts of each country to address climate change.

Fundamental to our use of renewable energies will be our ability to store that energy safely and affordably, balancing supply with demand.

Our storage of energy will mainly be in the form of batteries.

Battery [Ba] – source of electric power for powering a variety of electrical devices including flashlights, watches, computers, and automobiles. Batteries can be either primary (non-rechargeable) or secondary (rechargeable). There are also a variety of types of batteries including alkaline, lithium-ion, and solid-state.

Primary (non-rechargeable) battery [PB] – a battery whose chemical structure is not designed to be returned to its original charged state. Batteries in watches, key fobs, hearing aids, etc. are primary batteries (and often called button batteries). Most alkaline batteries are primary batteries.

Secondary (rechargeable) battery [SB] – a battery whose chemical structure is designed to be returned to its original charge state through a recharging process. A lithium-ion battery is the most common secondary battery.

Alkaline battery [AB] – a battery that stores energy in chemical form as a reaction between zinc metal and manganese dioxide. Alkaline batteries are the most dominant type of battery worldwide. Some alkaline batteries are designed, manufactured, and advertised to be secondary or rechargeable batteries.

Lithium-ion battery [LIB] – a rechargeable (secondary) battery in which lithium ions move from a negative electrode to a positive electrode when the energy is extracted. During the recharging process, the lithium ions move in reverse from the positive electrode to the negative electrode. Lithium-ion batteries are popular in use for portable electronics (smartphones, tablets, laptops, etc.), electric vehicles, and power tools such as drills, skill saws, lawn mowers, and the like.

Electric-vehicle battery [EVB] – a lithium-ion battery used in electric and hybrid electric vehicles. There are many variations of EVBs (i.e., lithium-ion battery for vehicles) including lithium-iron-phosphate (LFP) and nickel-cobalt-aluminum.

Solid-state battery [SSB] – a rechargeable (secondary) battery that utilizes solid electrodes and electrolytes instead of liquid and polymer gel electrodes and electrolytes like those used in current lithium-ion batteries. Because of the solid nature of the materials, solid-state batteries are much less susceptible to catching fire, leaks, and explosions. While right now expensive to make and still in their development, solid-state batteries hold the promises of:
- Higher energy densities
- Faster recharging
- Higher voltage
- Longer life

The Periodic Table of the Fourth Industrial Revolution

Communications Technologies

								ST	AV	DAT	LC	D						
								AD	eVTOL	W3D	Mv	Ro						
				3DB												XR	AR	VR
F				3DMS	SIS	STLF	C	3DR	PKC	PK	PrK	CE	CW	SiC	NFT	VRH	MR	MRH
FDM	R	3DB		IoAT	BA	GC		Bc	BN	BHK	M	I	Dapp	SCt	Hg	A	SM
DLP	SH			HoT	TMAI	AGI	SAAI	ASI	ANN	ML	SL	UL	DL	GA	DeFi	Pt	DAO
IoET	SG			RAI	LMAI										Ne	Bmm	NLP
3DP	IoT	IoCT	SC	AI	ANI												

													QRC	HaT	DST
						TA	Q	OC	N	Nt	DLT	CC	UPC	Brc	ET
			Bt	RFID	NFC	EMV	WiFi	LiFi	5G	6G	Ti	PA	GR	MC
CT		SS	SR	CV	S	L	Rdr	360C	SLAM	B	EVB	AB	PB
SsT	HT	CIC	CZ	CS	RE	PvS	MoSo	WF	GP	HFC	Pe	SSB	LiB	SB
E	GhG			EH	SP	CSP	WP	Hp	Bm	OP	Ba			
BTU	6E	GW	CN											

CHAPTER 2
COMMUNICATIONS TECHNOLOGIES

Communications technology [CT] – essential enabler of the movement of information from one place to another.

There exist both wired and wireless communications technologies. While wired communications technologies will remain important, the focus in the fourth industrial revolution will be on the continued development and use of wireless communications technologies. (Only wireless communications technologies are presented in the periodic table.)

Bluetooth [Bt] – a short-range wireless communications technology that has been standardized for communications between devices (fixed and mobile) over short distances, typically up to about 10 meters, 30 feet. Depending on the class of Bluetooth, lack of interference, and the connected devices themselves, range could be as much as 100 meters. Common personal devices like TV remote controls, wireless headphones, and game controllers are Class 2 devices and work effectively in the 10 meter range. (Class 1 devices have the greatest range.)

There are several variations of Bluetooth such as Bluetooth Classic and Bluetooth LE. Regarding the latter,

LE stands for *low energy*, and is designed for intermittent connectivity when not moving a lot of information.

Bluetooth comes standard on just about every personal computing device such as smartphones.

Radio frequency identification [RFID] – a communications technology that uses electromagnetic radio waves (sent by an RFID reader device) to identify and track RFID tags attached to objects. The reader device sends out a radio wave and the RFID chip responds by sending back digital data, which at a minimum is some sort of unique identification for the tag.

RFID has several variations, such as passive and active. A passive RFID has no internal power source and is effective up to a range of about 10 meters. Active RFID do have an internal power source and are effective up to about 100 meters.

RFID can also be categorized as either transponder or beacon. Transponder RFID activate and send information only when they receive a radio signal from a reader. So, transponder RFID remain dormant until they come into range of a radio signal. Beacon RFID continuously send a communication at some predetermined time interval. These are often used in real-time locating systems (RTLSs) and transmit not only the unique identification for the tag but also a location.

RFID range in size from about the size of a mailing stamp (basic passive RFID tag) to devices like those used on many toll roads.

Near-field communication [NFC] – a communications technology that is a subset of RFID but with limitations regarding data transmission speed, amount of data, and distance. Most all NFC applications are limited to a few centimeters, usually no more than 10, about 4 inches. The most well-known use of NFC is EMV.

EuroPay-Mastercard-Visa [EMV] – the communications technology for enabling the contactless reading of a chip on debit and credit cards. EuroPay, Mastercard, and Visa worked together to develop EMV and were the first to roll it out in the payments space. It is now standard for contactless cards.

Wireless fidelity [WiFi] – a suite of wireless network communications technologies and protocols that support local area networking for devices and Internet access. WiFi is the most commonly used network infrastructure in the world, especially for homes, offices, coffee shops, airports, etc. Effective range is typically about 75 feet but can range up to 500 feet, especially in some outdoor environments.

Light fidelity [LiFi] – an emerging communications technology that transmits data via light, as opposed to WiFi that uses radio frequency waves. LiFi has the possibility of transmitting data at extremely high speeds using visible light, ultraviolet, and infrared.

The promise of LiFi is substantial. It can transmit data in excess of 100 gigabits per second. The typical home WiFi unit can transmit data in the 1 gigabit range, and more realistically in the 100 to 200 megabit range. As well, because LiFi uses light transmission, it is not susceptible to interference from RF-based networks.

Still in its early stages of research and application, but definitely a technology to watch.

The Periodic Table of the Fourth Industrial Revolution

5G [5G] – the latest communications technology and set of protocols for broadband cellular networks. Released in 2018, 5G is an upgrade from its predecessor, 4G.

A broadband cellular network is a collection of data sending/receiving stations within a cell, a specific area of land. As you move from one cell to the next while travelling, your wireless device (car, phone, etc.) connects to the sending/receiving stations in a different cell. Because a cell has multiple sending/receiving stations, the broadband cellular network uses trilateration to measure your distance from each of station, thus being able to pinpoint your location.

Depending on the 5G coverage, download speeds can be in the 1 gigabit per second to 10 gigabits per second range.

6G [6G] – the next communications technology and set of protocols being developed for broadband cellular networks. 6G may very well usher in the age of mobile artificial intelligence, virtual reality, Internet of Things, and other fourth industrial revolution technologies, with no perceived latency.

Latency [Lt] – the amount of time is takes for a packet of data to move from one place to another. Latency is basically the time delay or lag you notice when accessing a website, uploading a photo, downloading a video, etc. As wireless communications technologies such as 5G and 6G become better and faster, the time delay or lag may not even be noticeable.

Tethering [Tt] – the sharing of an Internet connection with another device. Tethering is commonly associated with the term *hotspot*.

The Periodic Table of the Fourth Industrial Revolution

The Periodic Table of the Fourth Industrial Revolution

Sensing Technologies

																		ST	AV	DAT	LC	D	
																		AD	eVTOL	W30	Mv	Ro	
3DP	FDM	IoT	DLP	IoCT	IoET	SC	SG	AI	ANI		F	R	3DB		SH	IIoT	RAI	SIS	BA	TMAI	STLF	GC	AGI
																		3DMS	IoAT	LMAI	C	3DR	SAAI
																		PKC	Bc	ASI	PK	BN	ANN
																		PrK	BHK	ML	CE	M	SL
																		CW	I	UL	SiC	Dapp	DL
																		NFT	SCi	GA	XR	VRH	Hg
																		DeFi	Ne	AR	MR	A	Pt
																		Bmm	VR	MRH	SM	DAO	NLP

CT	Bt	TA	Q	QC	N	Nt	DLT	CC	EC	FC	DT	DST	HaT	QRC
SsT	RFID	NFC	EMV	WiFi	LiFi	SG	6G	Li	Ti	HFT	ET	Brc	UPC	
E	HT	SS	SR	CV	S	L	Rdr	360C	OCR	SLAM	B	MC	GR	PA
BTU	GhG	CIC	Cz	CS	RE	PvS	MoSo	WF	GP	HFC	Pe	PB	AB	EVB
	GE	GW	CN	EH	SP	CSP	WP	Hp	Bm	OP	Ba	SB	LiB	SSB

CHAPTER 3
SENSING TECHNOLOGIES

Sensing technology [SsT] – a technology that replicates the human senses, including sight, sound, touch, smell, taste, vestibular, and proprioception. The first 5 are commonly recognized. Vestibular includes movement and balance sense, which gives us information about where our head and body are in space. Proprioception addresses body awareness, which tells us about where our body parts are relative to each other.

Sensing technologies either (1) capture information, the role of input devices and/or (2) present information, the role of output devices. In that regard and in the broadest sense, a keyboard could be considered to be a sensing technology, as could a computer screen. These sorts of technologies (including mice, speakers, etc.) have been around for many years. The focus within the fourth industrial revolution will be those sensing technologies that sense data at the point of origin and/or those sensing technologies that further enhance our reception of information.

The broad categories of sensing technologies include:
- Hearing technology
- Computer vision (seeing)
- Haptic technology (feeling)
- Digital scent technology (smelling)

Hearing technology [HT] – a sensing technology that detects and may interpret sound, which could be the sound of a door closing, your voice, a simple thump, or even the sound termites make while eating the frame of your house.

Sound sensor [SS] – a microphone-based sensor that detects the presence of sound or noise, with the system using the sound sensor providing some sort of response. For example, a sharp loud noise outside might cause the security lights to come on. (Home security and monitoring is a popular application that uses sound sensors.)

You can choose from among a wide variety of microphone types (dynamic, carbon, ribbon, or condenser) and sizes, and you can program a sound sensor to control its sensitivity to sound.

Sound sensors differ from more advanced speech recognition which not only detect sound but also capture the sound at a certain quality to make sense of words.

Speech recognition (automatic speech recognition) [SR] – a sensing technology that detects, interprets, and responds to spoken words. In this case, the sound sensor is part of a complete system that includes AI to interpret words and usually sound creation to respond appropriately.

Things like Amazon Alexa, Google Assistant, and Siri rely on speech recognition technologies, as does the remote control for your TV.

Many such hearing technologies incorporate machine learning. The more you speak to a hearing technology, the more it learns to recognize not only your voice but also your tone, inflection, and use of certain words.

There are many implementations of speech recognition, for example elevators that respond to commands, and music recognition apps like Shazam, SoundHound, and Musixmatch.

Computer vision[CV] – a sensing technology that supports image recognition and understanding by a computer. Computer vision includes a wealth of technologies, each designed for a specific purpose.

Many computer vision technologies rely on sonar, Lidar, radar, or combination of the three. These all work on the same basic principle (adapted from the echolocation ability of animals and mammals); they measure the time it takes for a signal to reach an object, bounce off of that object, and return to the sending unit. They differ in the signal that each use.

Sonar (Sound Navigation and Ranging) [S] – a computer vision technology that measures the time it takes for a sound signal to reach an object, bounce off of that object, and return to the sending unit.

Lidar (Light Detection and Ranging) [L] – a computer vision technology that measures the time it takes for a light wave signal to reach an object, bounce off of that object, and return to the sending unit.

Radar (Radio Detection and Ranging) [Rdr] – a computer vision technology that measures the time it takes for a radio wave signal to reach an object, bounce off of that object, and return to the sending unit.

For a long time now, sonar has been used for "seeing" in the water. There are also several implementations in the medical realm, ultrasound and ultrasonic used to peek inside the human body.

Radar is popular for spotting aircraft in the sky (air traffic controllers) and determining the speed of moving vehicles (law enforcement).

Lidar is all the talk for autonomous vehicles. Lidar emits millions of laser light pulses to create a 3D map of the surrounding area. All autonomous vehicles use a combination of sonar, radar, and Lidar, to some varying degree.

As well, new releases of smartphones are incorporating Lidar. With Lidar, for example, you can map the exact layout of a room, capturing dimensions such as height, length, width, distance, and depth of the room and also objects such as furniture in the room.

360 camera (omnidirectional camera) [360C] – a computer vision technology that has a 360-degree seeing and capturing ability, thus giving it the ability to capture both still photos and videos of the entire surrounding area. 360 cameras are popularly used for surveillance and security, Google Street View, live events attended virtually, real estate, and virtual reality.

Regarding virtual reality, 360 cameras are popular because of your need to move your head and see in different directions. 360 cameras used for this purpose are often called *VR cameras*.

Optical character recognition [OCR] – the electronic conversion of typed, handwritten, or printed text into a computer-usable format. Many personal printers come with OCR, giving you the ability to scan a printed document and convert it into an editable format.

Simultaneous localization and mapping [SLAM] – a computer vision technology for environments that need the constructing and updating of an environment including the location of the "agent" within it. The iRobot Roomba uses SLAM to map a room, detect objects to avoid, and keep track of where it is in the room.

Biometrics [B] – literally means the measures of life, a combination of "bio" or life and "metrics" or measures of. Biometrics can be straightforward such as taking your blood pressure or in advanced applications like using your fingerprint for identification and really advanced applications in biomedical engineering, bio-technology, and neuro-technology. (The full spectrum of biometrics includes the capturing of many different types of information, such as a fingerprint or blood pressure. Fingerprint scanning relates to computer vision, while blood pressure relates to haptic technology. Thus, biometrics could also be included within the haptic technology grouping.)

Biometrics is an important aspect of other fourth industrial revolution applications such as 3D bioprinting to create synthetic skin grafts, and someday 3D-bioprinted fully artificial organs.

Motion capture [MC] – the process of detecting and recording the movement of objects, which can include people. Motion capture is commonly used in filmmaking and video game development to capture the movement of actors to create 2D and 3D animated renderings. It is also heavily used in the development of better safety and sports equipment such as helmets, safety harnesses, and the like.

Three important aspects of motion capture include gesture recognition, hand and finger tracking, and eye tracking.

Gesture recognition [GR] – a technology with the goal of recognizing human gestures, including the shrugging of shoulders, changes in facial expressions due to emotional changes, in-the-air swiping by the hand for touchless interfaces, body movements while using virtual weapons, and so on.

Hand and finger tracking [HFT] – a type of motion capture for detecting and understanding finger and hand movements. A popular application of hand and finger tracking is the interpretation of sign language.

Eye tracking [ET] – the process of measuring the point of gaze, i.e., where someone is looking. Eye-tracking technologies enable marketers to detect what captures a consumer's attention, in what order a consumer reads information on a box, and so on.

An important subset of computer vision are data in a machine-readable, standardized format. The main ones include barcode and its 2 major variations, UPC and QR code.

Barcode [Brc] – a method of representing data in a machine-readable, standardized format. Popular barcode types include UPC and QR code.

Universal product code [UPC] – a type of barcode that uses vertical bars, with the size of the bars and the distance between them determining the number. These are referred to as linear or 1-dimensional barcodes. A UPC has 12 numbers, with the most common application being a product number in an inventory system. A UPC is the same code for all of the same product.

Quick response code [QRC] – a matrix barcode and referred to as a 2D (2-dimensional) barcode. QR codes can contain much more information than a UPC. There are many different configurations of QR codes including Aztec Code, SnapTag, and SPARQCode.

Haptic technology [HaT] – a sensing technology that (1) creates the sense of touch by applying force and motion (including vibration) to a person and/or (2) detects and interprets force or motion from a person or object. Thus, haptic technologies are a two-way street; the computer is sending you "feeling" through haptic interfaces and you are sending the computer "feeling" in the form of your motions, actions, etc.

Broadly, the three categories of haptic technologies are (including a few examples):
1. When you "feel" the computer – game controllers and chairs that vibrate to provide a more immersive gaming experience such as feeling an explosion, and driver-assistance technologies in automobiles that provide vibration in response to irregular or unsafe driving.
2. When the computer "feels" you – simple technologies like mice, touchpads, joysticks, and hand controllers for extended reality applications, and electronic headsets that adjust their field of vision to your moving your head in different directions.
3. When the computer "feels" other things – vibration/motion sensors (for example, placed on equipment to determine when it may be coming out of balance), pressure sensors (to determine, for example, weight change), and tilt sensors (for detecting a change in angular orientation).

Digital scent technology (olfactory technology) [DST] – a sensing technology that can detect natural scents, smells, molecules, and particles in the air. Smelling technologies attempt to replicate the human olfactory system, smell.

Smelling technologies can detect typical odors that humans can smell, as an example perfume, flowers, or dirty socks. They can also detect odors that humans cannot smell, as an example carbon monoxide.

As with many sensor technologies, smelling technologies can also produce synthetic odors, an emerging sensory output for many extended reality (XR) applications.

An interesting fourth industrial revolution technology that relies on many of the sensing technologies is that of a digital twin.

Digital twin [DT] – a real-time virtual counterpart of a physical object. Take the wind turbine as an example. To a wind turbine you can add IoT sensors all over it as well as the platform and surrounding ground. The IoT sensors can measure stress, motion/vibration, speed of the wind, stability of the platform, changes in the soil structure around the turbine platform, and so on. All those taken together give you the ability to monitor the turbine and turbine location (i.e., the turbine digital twin) without needing to be there physically for an inspection.

NOTE: THERE IS A SENSOR FOR EVERYTHING...

I specifically included sound sensors in this section and eluded to others. Indeed, if you need to measure something, there is a sensor for it. Some of those would include:

- GPS (location), accelerometer (speed), gyroscope (direction)
- Proximity/Ultrasonic (how close is an object), motion, infrared, tracking, vibration
- Barometric pressure, gas pressure, weight (pressure)
- Temperature and humidity
- Water level, water flow, moisture
- Magnetic detection and metal touch
- Chemical, gas, carbon monoxide/dioxide
- Smoke, flames
- Light
- Sound
- Water quality (e.g., pH, chlorine-related)
- Image recognition, color recognition

I didn't list each of these individually in the periodic table. There are hundreds, which would be a periodic table all to itself.

Technology Architectures

															ST	AV	DAT	LC	D								
															AD	eVTOL	W3D	Mv	Ro								
3DP	FDM	F						C	PKC	PK	PvK	CE	CW	StC	NFT	XR	VRH	Hg	DeFi	Ne
IoT	DLP	R	3DB						Bc	BN	BHK	M	I	Dapp	SCt	AR	MR	A	Ft	Bmm
IoCT	IoET	SH	3DMS	STLF	SIS	3DR														
SC	SG	IIoT	IoAT	GC	BA															
AI	ANI	RAI	LMAI	AGI	TMAI	SAAI	ASI	ANN	ML	SL	UL	DL	GA	VR	MRH	SM	DAO	NLP

CT	Bt	RFID	NFC	Q	QC	N	Nt	DLT	CC	EC	FC	DT	DST	HaT	QRC
SsT	HT	SS	SR	EMV	WiFi	LiFi	5G	6G	Li	Tt	HFT	ET	Brc	UPC
E	GhG	CIC	CZ	CV	S	L	Rdr	360C	OCR	SLAM	B	MC	GR	PA
BTU	GE	GW	CN	CS	RE	PvS	MoSo	WF	GP	HFC	Pe	PB	AB	EVB
				EH	SP	CSP	WP	Hp	Bm	OP	Ba	SB	LIB	SSB

CHAPTER 4
TECHNOLOGY ARCHITECTURES

Technology architecture [TA] – refers to (1) the internal workings and structure of a computer itself or (2) the physical configuration of a technology system within a physical area.

The former includes quantum computing and nanotechnology. The latter includes distributed ledger technology, cloud computing, edge computing, and fog computing.

Classical computing technologies, our current paradigm, work in a form called binary. In binary, there are only 2 basic representations of information or data, either a 0 or a 1. These are referred to as a bit, which stands for <u>bi</u>nary <u>digit</u>. All types of information – numbers, letters, special symbols, etc. – are represented as a series of bits inside a computer. For example, the letter P is 01010000.

Binary is an implementation of basic electronics, either a gate is open or closed, either electricity is present or not, and so on. It is analogous to a coin flip (either heads or tails), a light being on or off, true or false, etc.

Since the inception of classical computing technologies in the 1940s, the focus has been on making smaller, better, faster, and more affordable binary-based technologies.

This has run the gambit from vacuum tubes to transistors to integrated circuits to very-large scale integration (VLSI).

Research is now underway to shift away from classical computing technologies and into the realm of quantum computing and using qubits instead of bits.

Qubit [Q] – the basic unit for representing information in quantum computing. Qubit stands for <u>q</u>uantum <u>bi</u>nary digi<u>t</u>. Qubits exhibit numerous unique characteristics, one of which is superposition, the ability of a qubit to be both 0 and 1 at the same time. (Recall that classical computing technology bits can be either 0 or 1, but not both at the same time.)

As such, qubits enable scaling of speed. For example, adding all possible combinations of 2 classical bits would involve 4 computations: $0 + 0 = 0, 0 + 1 = 1, 1 + 0 = 1$, and $1 + 1 = 10$. (10 is not a typo; there is no 2 in binary.) Because a qubit can be both 0 and 1 at the same time, adding all possible combinations of 2 qubits is a <u>single</u> computation. Thus, scaling speed when adding more qubits is 2^n, where *n* is the number of qubits used.

Quantum computing [QC] – an alternative to classical computing technology that offers unbelievable speed because it performs computations on information in states such as superposition.

In October 2019, Google announced that it had achieved *quantum supremacy*. Its quantum computer solved a highly-complex mathematical computation in 3 minutes and 20 seconds that would have taken the world's fastest supercomputer (based on classical binary-based computing technology) over 10,000 years to solve. [See https://www.nytimes.com/2019/10/23/technology/quantum-computing-google.html] This achievement in speed was due to the scaling achieved by using successively more qubits in the computation.

While quantum computing focuses on speed, nanotechnology focuses on physical size.

Nano [N] – one-billionth. 1 nanometer is one-billionth of a meter. There are just over 25 million nanometers in an inch.
- A human hair is about 100,000 nanometers in width.
- A sheet of paper is about 100,000 nanometers thick.
- Your fingernails grow about 1 nanometer per second.
- You hair grows at about the equivalent speed.
- The popular comparative scale – If a marble were a nanometer, then one meter would be the size of the earth.

Nanotechnology [Nt] – the manipulation of material with at least one dimension of size between 1 and 100 nanometers. Something that is 100 nanometers in width is 1,000 times smaller than a human hair.

The possibilities of nanotechnology can go even smaller. Think about being to build things on a molecule by molecule basis, or an atom by atom basis. It's theoretically possible to build a complete computer using nanotechnology – RAM, CPU, battery power supply, etc. – that would be about the width of a human hair.

Distributed ledger technology [DLT] – a technology environment in which all "nodes" on a distributed network (1) have an accurate and up-to-date copy of the information, (2) the software for processing transactions and maintaining the information, and (3) approve any changes to and/or uses of the information.

The Periodic Table of the Fourth Industrial Revolution

The vast majority of technology systems today are based on a centralized ledger technology, with one "master" set of information stored in a centralized database, or ledger, and one central authority/organization that oversees the quality, updating, and use of the centralized database/ledger. The U.S. banking system is based on a centralized ledger technology concept.

When you deposit a personal check from a friend or family member, it takes 2 to 5 days for the check to clear and for you to have access to the funds. Your bank has no way of knowing or authenticating that the other account has sufficient funds to cover the check. The check and the movement of money have to be processed centrally.

Using our financial system as an example for a distributed ledger technology, all banks (and savings and loans, credit unions, credit card providers, brokerages, etc.) would be a node on the distributed network. Each would have an up-to-date and accurate copy of all financial transaction information. Each would have the ability to process transactions, and each would validate every transaction (originating at its location or elsewhere).

When a node initiates a transaction, such as your depositing a personal check from a family member, that node would send the details of the transaction to all the other nodes. Those nodes would determine the validity of the transaction using their respective copy of the master set of information. If valid, the nodes would approve the transaction. Then, all nodes would execute the transaction and update their copy of the master information. (You, also, would have instant access to the funds of the deposited check.)

That is the concept of a distributed ledger technology.

Distributed ledger technology is a prerequisite to the use of blockchain (Bc). That is to say, blockchain is only possible as a distributed ledger technology.

Cloud computing [CC] – storing and accessing of information, software, and/or processing on the Internet. Popular cloud providers include Amazon Web Services (AWS), Microsoft Azure, Google Cloud, Rackspace Cloud, and VMWare. (There are many, many others.)

Edge computing [EC] – processing (and perhaps storage) of information very close – if not on – the logical edge of the network where the information is captured at its source.

Fog computing [FC] – a hybrid of cloud and edge computing in which information is processed (and, again, perhaps stored) on a local area network within the enterprise.

Cloud, edge, and fog computing in essence deal with (1) where the information is captured and (2) where the information is process and stored. Within cloud computing, there exists the greatest distance between the capturing of the information and the storage/processing of the information (i.e., the Internet). Within edge computing, there exists the shortest distance between the capturing of the information and the storage/processing of the information. Fog computing (a term originally coined by Cisco) is somewhere in between.

The choice of which to use is completely dependent on the situation and the information, storage, and processing requirements.

The Internet Of Things

							ST	AV	DAT	LC	D				
							AD	eVTOL	W3D	Mv	Ro				

3DP	FDM	F	3DB										XR	AR	VR
													VRH	MR	MRH
IoT	DLP	R	3DMS	STLF	SiS	C	PKC	PK	PrK	CE	CW	SiC	NFT	Hg	A	SM
IoCT	IoET	SH	IoaT	GC	BA	3DR	Bc	BN	BHK	M	I	Dapp	SCt	DeFi	Fr	DAO
SC	SG	IIoT	LMAI	AGI	TMAI	SAAI	ASI	ANN	ML	SL	UL	DL	GA	Ne	Bmm	NLP
AI	ANI	RAI														

CT	Bt	Q	TA	QC	N	Nt	DLT	CC	EC	FC	DT	DST	HaT	QRC
SsT	HT	RFID	NFC	EMV	WiFi	LiFi	5G	6G	Li	Tt	HFT	ET	Brc	UPC
E	GNG	SS	SR	CV	S	L	Rdr	360C	OCR	SLAM	B	MC	GR	PA
BTU	GE	CIC	CZ	RE	PvS	MoSo	WF	GP	HFC	Pe	PB	AB	EVB
		CN	GW	SP	CSP	WP	Hp	Bm	OP	Ba	SB	LiB	SSB

CHAPTER 5
THE INTERNET OF THINGS

Internet of Things [IoT] – a network of inter-connected objects that collect, process, and exchange data. Those objects can include computers, smartphones, sensors, cars, and so on.

IoT already exists, and like other fourth industrial revolution technologies, is arriving in stages. We have basic IoCT (Stage #1), we're deep in the middle of IoET (Stage #2), and we're moving into IoAT (Stage #3).

Internet of Computer Things [IoCT] – began in the third industrial revolution with the connecting of computers and computer-related things to the Internet, and communicating with each other and sharing information, data files, photos, etc.

You can use your phone to share photos with your computer. You can use your laptop or tablet to locate your phone. You can easily and dynamically create, edit, and share files among your computer devices using services like iCloud, Dropbox, and Google Docs & Drive.

IoCT is here and in full swing.

Internet of Electronic Things [IoET] – the connecting of all things electronic to the Internet. Think here about household electronic devices such as washing machines, dryers, refrigerators, heating and air conditioning systems, garage doors, water sprinkler systems, stereos, and TVs.

Things like Google Nest (Learning Thermostat, Doorbell, and Protect) and Ring (signature video-based doorbell systems and other home-oriented systems) are an important part of IoET. You can even include Amazon Alexa and Google Assistant in IoET, both of which can answer questions, control various parts of your home, and generally make you feel like you never have to leave the comfort of your couch (or touch a remote control).

Many IoET applications fall into the realms of smart homes, smart cities, smart grids, and the industrial Internet of Things.

Smart Home (Home Automation or Domotics) [SH] – building automation systems and intelligence for your home. You can now use apps to control your lights, garage door, and setting for your water sprinkler system.

Smart city [SC] – technologically-based city environments that use a wealth of technologies to gather data and use that data to more efficiently deliver services related to such things as utilities, transportation, crime detection, medical care, waste removal, snow plowing, and weather-related warning signals.

Smart grid [SG] – when a city applies IoT to improve the delivery, efficiency, and reliability of utilities. This can include electricity, gas, and water.

Industrial Internet of Things [IIoT] – the use of IoT in commercial and industrial settings as opposed to consumer-oriented IoT like Google Nest and Ring. Major manufacturing operations use IIoT to monitor manufacturing equipment. IoT sensors can measure vibration, for example, to detect early warning signals that equipment may be coming out of alignment and in need to repair/maintenance. Smart cities are often included in IIoT.

Internet of All Things [IoAT] – the connecting of all things non-electronic to the Internet. Early examples of IoAT include toothbrushes that monitor your brushing effectiveness, water bottles that measure fluid intake and fluid quality, furniture such as beds that monitor your sleep, and even clothing (e.g., shoes with embedded sensors that measure your speed and distance travelled.)

IoAT is the ultimate goal of IoT, everything connected to the Internet… collecting, processing, and exchanging data to make better decisions.

The Periodic Table of the Fourth Industrial Revolution

Artificial Intelligence

3DP	FDM	F				ST	AV	DAT	LC	D				XR	AR	VR
IoT	DLP	R	3DB			AD	eVTOL	W30	Mv	Ro				VRH	MR	MRH
IoCT	IoET	SH	3DMS	STLF	SIS	C	PKC	PK	PrK	CE	CW	SiC	NFT	Hg	A	SM
SC	SG	HoT	IoAT	GC	BA	3DR	Bc	BN	BHK	M	I	Dapp	SCI	DeFi	Ft	DAO
AI	ANI	RAI	LMAI	AGI	TMAI	SAAI	ASI	ANN	ML	SL	UL	DL	GA	Ne	Bmm	NLP

CT	BR			TA	Q	QC	N	Nt	DLT	CC	EC	FC	DT	DST	HaT	QRC
SsT	HT	SS			RFID	NFC	EMV	WiFi	LiFi	5G	6G	Ti	HFT	ET	Brc	UPC
E	GhG	CIC	SR	CV	S	L	Rdr	360C	OCR	SLAM	B	MC	GR	PA
BTU	GE	GW	CZ	CS	PvS	MoSo	WF	GP	HFC	Pe	PB	AB	EVB
			CN	EH	SP	CSP	WP	Hp	Bm	OP	Ba	SB	LIB	SSB

40

CHAPTER 6
ARTIFICIAL INTELLIGENCE

Artificial intelligence [AI] – intelligence demonstrated by machines or technology. As such, the goal of AI is to create technologies that mimic human intellectual tasks such as learning, problem solving, and even social interaction.

AI has formally been around since the mid 20th century, marked by significant funding from both the private and public sector, early commercial successes with expert systems, numerous failures, grandiose ideas that never came to fruition, and of course successes in numerous realms such as autonomous vehicles, facial recognition, ferreting out fraudulent insurance claims, optimizing supply chain management routes, matching people to content on social media platforms, and even the Roomba, the popular iRobot vacuum cleaner.

There are 7 (somewhat overlapping) categories of artificial intelligence:
1. Artificial narrow intelligence
2. Reactive artificial intelligence
3. Limited-memory artificial intelligence
4. Artificial general intelligence
5. Theory of mind artificial intelligence
6. Self-aware artificial intelligence
7. Artificial superintelligence

Artificial narrow intelligence [ANI] – an AI that performs one and <u>only</u> one task. This is the dominate category of AI in terms of successful applications. Artificial narrow intelligence includes the likes of Netflix's recommendation engine, car engine diagnostics software, chatbots, virtual assistants, spam filters, image recognition, handwriting recognition, medical diagnosis and prescription, and autonomous vehicles.

Narrow AIs are programmed for one specific situation. You can build a narrow AI to play chess, but it will not be able to play checkers. A narrow AI does one and <u>only</u> one thing.

Reactive artificial intelligence [RAI] – a subset of artificial narrow intelligence in which the AI is programmed to provide a predictable outcome based on the inputs it receives.
- The same set of inputs will always produce the same outcome. Always.
- It will always respond to the same situation (inputs) in exactly the same way each and every time.
- It cannot handle new inputs, nor can it adapt itself to new situations.
- It cannot learn.

Reactive AI is commonly implemented as a series of if-then rules to work through a set of inputs to determine what action to take or recommend. Think of driving through an intersection. Is the light green? If yes, then procced through. If the answer is no, then evaluate the next rule: Is the light red? If yes, then stop. If no, then the light must be yellow (evaluate the next rule). Do you have time to stop? If yes, then stop. If no, then accelerate and proceed through the intersection.

Limited-memory artificial intelligence [LMAI] – a subset of artificial narrow intelligence in which the AI learns from historical/past data and real-time feedback to make better decisions. So, the more a limited-memory AI is used, continually learning, the better it gets at making its decision.

Within the dominant category of narrow AI, limited-memory AI is the most dominant subset. This is really where the vast majority of successful AI applications exist.

Limited-memory artificial intelligence:
- Most often implemented via an artificial neural network (ANN).
- Can adapt over time to changing conditions and inputs.
- Exhibits machine learning (ML), which can become deep learning (DL).
- Exhibits the ability to learn.

Artificial general intelligence [AGI] – an AI that can take what it knows and has learned for one situation and apply it to a new situation. These AI will function completely like a person, intellectually. They will be able to build competencies in situations without being explicitly programmed for them. They will be able to generalize across domains, adapting their current knowledge to new situations.

Theoretically possible but not yet a reality. Many people don't believe we'll ever be able to create a general artificial intelligence. We certainly cannot with our current technologies. But there is much research in this space.

Theory of mind artificial intelligence [TMAI] – an AI that interacts socially and can discern and respond appropriately to people's emotions, beliefs, thoughts, expectations, and facial expressions. Theory of mind AI takes AI beyond intellectual tasks and into the realm of "understanding" people.

There are a couple of limited successes in this area. See, for example, Sophia (a humanoid robot designed by Hanson Robotics) and Kismet (a robot head designed by Cynthia Breazeal at MIT).

Self-aware artificial intelligence [SAAI] – an AI that is for all practical purposes a person. It will have emotions and be aware of its emotions. It will be sentient. It will have a conscience. It will understand the need to procreate. It will struggle with ethical decisions. It will probably even act irrationally at times, just as we do, and later apologize.

Purely theoretical.

Artificial superintelligence [ASI] – the most far-reaching view of the potential of AI. An artificial superintelligence would be the most intelligent entity on the earth, making better and faster decisions than humans in all aspects of life. Artificial superintelligence is popularly referred to as *singularity*, the point at which machines become smarter than the entire human race.

Purely theoretical.

Artificial neural network [ANN] – a software tool with workings patterned after the human brain, including things like axons, dendrites, neurons, and synaptic connections. Often called just neural networks, these exhibit the characteristic of learning, continually refining the way they work to arrive at increasingly better outcomes with each use.

Neural networks are the most popular form of limited-memory artificial intelligence.

Neural networks have 3 layers:
1. Input layer – for accepting the inputs for a given situation
2. Hidden layer(s) – the layer(s) of neurons that, within each, have constantly changing mathematical formulas. These take in the inputs, mathematically manipulate them, and send the results to the output layer.
3. Output layer – the layer that accepts information from the hidden layers and then decides – based on some threshold value – what answer to provide.

Based on the correctness of response, a neural network changes the manner in which the hidden layer(s) works and how those neurons mathematically manipulate the inputs from the input layer. Thus, a neural network changes its responses and also learns to get better with each use.

Neural networks exhibit machine learning. They can learn via supervised learning or unsupervised learning. They can exhibit deep learning.

Machine learning [ML] – the ability of an AI to learn from data by changing the way it works and thus the answer at which it arrives in subsequent uses. Machine learning is commonly associated with neural networks. Based on a given set of inputs and what the correct answer is, the neural network adjusts the workings of its hidden layer(s), thus arriving at a better answer the next time it receives inputs.

This is the concept of training. This training or learning can be supervised or unsupervised.

Supervised learning [SL] – in artificial intelligence, is a training process in which the inputs are specified or labeled, and the correct answer is also provided (labeled). If the neural network didn't get the right answer, it adjusts its hidden layer(s) appropriately to get closer to getting the right answer the next time. If the neural network did get the right answer, it adjusts its hidden layer(s) appropriately to reinforce its learning.

Unsupervised learning [UL] – in artificial intelligence, is a training process in which a goal is provided to the neural network and the neural network chooses which inputs to use to best achieve the goal. This is similar to an optimization function. The neural network is told what to optimize, and it then determines – through repeated trials – what inputs to use to achieve the highest or best optimization. Unsupervised learning often uses huge amounts of unstructured data to answer very complex questions.

Deep learning [DL] – in artificial intelligence, is a subset of machine learning and the ability of an AI to get really, really good at making a decision, often in domains involving massive amounts of unstructured data. Deep learning usually involves unsupervised learning and is achieved by increasing the number of hidden layers in a neural network. Thus, all deep learning is machine learning, but not all machine learning is deep learning.

Deep learning can create a more effective AI. But, deep learning does require much more training and a longer learning process.

In thinking about machine learning, deep learning, supervised learning, and unsupervised learning, consider cats and dogs. Using supervised learning, a machine learning-based AI could get very good at distinguishing between a cat and a dog. However, for an AI to become effective at distinguishing among the some 300+ breeds of dogs, it would require deep learning via unsupervised learning.

Genetic algorithm [GA] – a set of techniques modeled after biologically-inspired intelligences that primarily include the following:
- Selection – choosing a better outcome over a poorer outcome
- Crossover – combining 2 good outcomes to see if a better outcome can be achieved
- Mutation – randomly trying something new to determine if a better outcome can be achieved

Neuroevolution [Ne] – the use of genetic algorithms within an unsupervised learning context. With neuroevolution, optimization strategies such as selection, crossover, and mutation become the primary drivers for how a neural network learns, without specifying the inputs to be used.

Biomimicry [Bmm] – the simulation and use of elements of nature to solve human problems. For example, we've learned much from hiving insects to determine how to better build things like buildings and homes. We've studied forager ants to determine how to better design supply management activities. We've studied birds to build better the aerodynamic nature of airplanes. We've studied termite mounds to better understand air flow and ventilation.

Natural language processing [NLP] – the use of artificial intelligence (AI) to interpret the human language, in both audio and written forms. Natural language processing is an important element of speech recognition.

The Periodic Table of the Fourth Industrial Revolution

Blockchain

3DP						ST	AV	DAT	LC	D				XR	AR	VR
IoT	FDM	F	3DB			AD	eVTOL	W30	Mv	Ro				VRH	MR	MRH
IoCT	DLP	R	3DMS	STLF	SIS	C	PKC	PK	PrK	CE	CW	SiC	NFT	Hg	A	SM
SC	IoET	SH	IoAT	GC	BA	3DR	Bc	BN	BHK	M	–	Dapp	SCI	DeFi	Fi	DAO
AI	ANI	RAI	LMAI	AGI	TMAI	SAAI	ASI	ANN	ML	SL	UL	DL	GA	Ne	Bmm	NLP

						TA	Q	QC	N	NI	DLT	CC	EC	DT				QRC
						RFID	NFC	EMV	WiFi	LiFi	5G	6G	Li	HFT	Tt	B	HFC	UPC
CT	Bt					SS	SR	CV	S	L	Rdr	360C	OCR	B	SLAM	Pe		PA
SsT	HT					CIC	CZ	CS	RE	PvS	MoSo	WF	GP	Pe	HFC	Ba	AB	EVB
E	GhG					GW	CN	EH	SP	CSP	WP	Hp	Bm	Ba	OP		LIB	SSB
BTU	GE																SB	

CHAPTER 7
BLOCKCHAIN

Blockchain [Bc] – an implementation of distributed ledger technology in which information is stored in the form of blocks, with each block holding multiple records or transactions, and with the blocks linked or chained together via a unique key or hash. Blockchain is only possible as an implementation of distributed ledger technology.

Blockchain node [BN] – the participants in a blockchain network who verify transactions, maintain the integrity of the network, keep an up-to-date copy of the distributed ledger, i.e., the blockchain, and agree on the updates to and use of blockchain information.

Because blockchain is an implementation of distributed ledger technology, there is no central authority for maintaining the integrity of the information, nor is there a single central repository of the information. Instead, each blockchain node maintains an up-to-date copy of the information. When a transaction is "announced" by a node, all other nodes verify the transaction by evaluating their copy of the information. If there is agreement within the network that the transaction is valid, then the transaction is executed by all nodes, with the nodes updating their respective copies of the blockchain.

Block hash or key [BHK] – a unique 256-bit 64-character series that uniquely identifies the contents of a block in a blockchain. Once a block is full, the block hash or key is created and added to the end of the block. That block is then permanently written to the blockchain by all blockchain nodes. The hash or key for that block also becomes the first entry into the next block. Thus, chaining the blocks together, both forward and backward.

Mining [M] – the process of using all the transactions in a block to create the unique hash or key for the block. In the popular press, mining is most often associated with the mining of cryptocurrency. However, all blockchain systems use mining to chain the blocks together via a unique hash or key.

Immutability [I] – a characteristic of a blockchain in which, once a transaction is approved and written to the blockchain, it cannot be altered or changed in any way. If that were to happen, the block hash or key for that block would have to change and no longer match the beginning entry for the next block. This effectively breaks the chain and creates a system error.

Decentralized app (decentralized application) [Dapp] – the software applications that run on a blockchain. Software applications that operate within a typical centralized architecture model are commonly referred to as apps and application software. Because the software and information are decentralized in a blockchain environment, the new term for software on a blockchain is Dapp. Dapps include smart contracts and decentralized finance.

Smart contract [SCt] – a piece of software that runs on a blockchain and automates processes, tasks, and the movement of money as certain conditions are met. Within a smart contract, blockchain nodes have the responsibility to verify that the conditions have been met.

In the building of a home, for example, the electrical subcontractor would provide all the wiring and electrical considerations. Once done, the county inspector, general contractor, and home owner would act as blockchain nodes and approve the work. Once all approve the work, the electrical subcontractor would immediately be paid.

Decentralized finance [DeFi] – an emerging financial technology based on distributed ledge technology and blockchain that removes the control, authority, and costs of financial intermediaries such as banks, brokerages, savings and loans, credit unions, etc.

Of particular interest is the use of DeFi to reach the 1.7 billion adults worldwide who are un-banked, meaning that they do not have an account with a financial institution. Many of these unbanked people cannot afford a traditional account because of the associated fees and required minimum monthly balances. They can use DeFi to directly move cryptocurrency to another individual, without the need for a traditional financial intermediary.

Fintech [Ft] – the integration of technologies – like blockchain – that improves the delivery, efficiency, and cost of basic financial services. Obviously, DeFi falls within the broader category of fintech.

Decentralized autonomous organization [DAO] – an Internet community with a shared bank account. [See https://www.cnbc.com/2021/10/25/what-are-daos-what-to-know-about-the-next-big-trend-in-crypto.html] A DAO is basically a group of people who decide to pool their money for any number of purposes and establish a set of rules (i.e., smart contract) by which the group will operate.

Creating a DAO is rather like forming a partnership with several of your friends. You pool your money into a single account, and then you establish the rules by which your partnership will operate. What makes DAOs particularly appealing is that you automate your operating rules in the form of a smart contract on a blockchain.

The Periodic Table of the Fourth Industrial Revolution

The Periodic Table of the Fourth Industrial Revolution

Cryptocurrency

							ST	AV	DAT	LC	D						XR	AR	VR
			3DB				AD	eVTDL	W30	Mv	Ro						VRH	MR	MRH
3DP	FDM		3DMS	STLF	SiS		C	PKC	PK	PrK	CE	CW	StC	NFT			Hg	A	SM
IoT	DLP	F	IoAT	GC	BA		3DR	Bc	BN	BHK	M	I	Dapp	SCt			DeFi	Ft	DAO
IoCT	IoET	R	LMAI	AGI	TMAI		SAAI	ASI	ANN	ML	SL	UL	DL	GA			Ne	Bmm	NLP
SC	SG	SH																	
AI	ANI	IIoT																	
		RAI																	

TA	Q	QC	N	Nt	DLT	CC	EC	FC	DT	DST	HaT	QRC
Bt	RFID	NFC	WiFi	LiFi	5G	6G	Li	Tt	HFT	ET	Brc	UPC
SsT	HT	SS	SR	L	Rdr	360C	OCR	SLAM	B	MC	GR	PA
E	GNG	CIC	CZ	PvS	MoSo	WF	GP	HFC	Pe	PB	AB	EVB
BTU	GE	GW	CN	CSP	WP	Hp	Bm	OP	Ba	SB	LIB	SSB
CT			CV									
			CS									
			EH									

CHAPTER 8
CRYPTOCURRENCY

Cryptocurrency [C] – a currency in digital or electronic form, with no physical equivalent. Cryptocurrency currency news is dominated by the likes of Bitcoin and ETH (short for Ether). All cryptocurrency is based on blockchain, and thus not possible without blockchain.

All cryptocurrency use cryptography to secure, validate, and authenticate the use of cryptocurrency in transactions. That's why you see "crypto" in front of currency.

Public key cryptography (asymmetric encryption) [PKC] – the type of cryptography used to secure, validate, and authenticate the use of cryptocurrency in transactions. (PKC has many other applications beyond cryptocurrency.) Public key cryptography makes use of pairings of public keys and private keys.

Public key [PK] – a key on a blockchain that is publicly known and is used mainly for purposes of identification. In the crypto world, someone has to know your public key in order to send you crypto. Think of your public key as your bank account and routing number.

Private key [PrK] – a key on a blockchain that gives you the sole and exclusive right to use cryptocurrency you own.

So, when you buy cryptocurrency (or a friend or customer sends it to you), you receive a private key for it, giving you the exclusive right to use it, send it to someone else, trade it for another cryptocurrency, etc. To keep your cryptocurrency safe, you need to securely store your private keys in a cryptocurrency wallet or on a cryptocurrency exchange.

Cryptocurrency exchange [CE] – a marketplace in which you can buy, sell, and hold cryptocurrency. Popular cryptocurrency exchanges include Coinbase, Exodus, Binance, Kraken, Gemini, and Crypto. Think of a cryptocurrency exchange as your bank holding your bank accounts, and thus your money. All cryptocurrency exchanges offer wallet capabilities.

Cryptocurrency wallet [CW] – a piece of software that allows you to store and manage your cryptocurrency.

There are many different types of cryptocurrency wallets such as custodial and non-custodial. Custodial wallets are offered by cryptocurrency exchanges. In this way, a custodial wallet is similar to your bank account, which is managed by your bank. A non-custodial wallet gives you complete control of your wallet and cryptocurrency. You store and manage your wallet and cryptocurrency on your own computer, independent of an exchange.

There are also hot and cold cryptocurrency wallets. A hot cryptocurrency wallet is one which is connected to the Internet, and thus readily available for use but also susceptible to hacks and thefts. A cold cryptocurrency wallet is not usually connected to the Internet. Most cold wallets are like thumb drives. You store your cryptocurrency on the thumb drive and then disconnect it from your computer (and Internet), eliminating the possibility of online theft.

Stablecoin [StC] – a type of cryptocurrency that is pegged or tied to an external asset like the U.S. dollar or gold. This helps stabilize the price of the stablecoin. The most popular cryptocurrencies like Bitcoin and ETH are not stablecoins, and thus have wild variations in price based on speculation.

Popular stablecoins include Tether, DAI, Binance USD, and Digix Gold Token.

Non-fungible token [NFT] – a digital certificate representing ownership. Most NFTs are found in the collectible world, representing ownership of photos, art, videos, and the like. They can also represent ownership of land (real and virtual), automobiles, homes, etc. When you buy an NFT, you are buying a certificate of ownership of something.

The Periodic Table of the Fourth Industrial Revolution

3D Printing

3DP	FDM	F															ST	AV	DAT	LC	D						XR	AR	VR
IoT	DLP	R	3DB														AD	evTOL	W30	Mv	Ro						VRH	MR	MRH
IoCT	IoET	SH	3DMS	SIS	STLF													C	PKC	PK	PrK	CE	CW	SiC	NFT		Hg	A	SM
SC	SG	IIoT	IoAT	BA	GC													3DR	Bc	BN	BHK	M	I	Dapp	SCi		DeFi	R	DAO
AI	ANI	RAJ	LMAJ	TMAJ	AGI													SAAI	ASI	ANN	ML	SL	UL	DL	GA		Ne	Bmm	NLP

TA	Q	QC	N	Nt	DLT	CC	EC	FC	DT																		DST	HaT	QRC
RFID	NFC	EMV	WiFi	LiFi	SG	6G	Lt	Ti	HFT																		ET	Brc	UPC
SS	SR	CV	S	L	Rdr	360C	OCR	SLAM	B																		MC	GR	PA
CIC	CZ	CS	RE	PvS	MoSo	WF	GP	HFC	Pe																		PB	AB	EVB
GW	CN	EH	SP	CSP	WP	Hp	Bm	OP	Ba																		SB	LiB	SSB

Bt	HT	GhG	GE
CT	SsT	E	BTU

CHAPTER 9
3D PRINTING

3D printing [3DP] – the construction of a 3-dimensional object by typically adding (i.e., printing) layer upon layer of liquid material until the object has been completely printed or constructed. The material is initially heated to a necessary liquid state and then printed. Once a layer cools, more material is heated and printed on top of the previous layer.

The most common types of 3D printers are fused deposition modeling and digital light processing. The former uses material called filament, and the latter uses material called resin.

Fused deposition modeling [FDM] – a 3D printing process that uses a spool of solid material, called filament. FDM 3D printers heat the filament and print it through a heated printer head called an extruder. FDM 3D printers are the most common. As they work with material called filament, FDM 3D printers are also referred to as filament 3D printers.

Filament [F] – a spool of solid material used by an FDM 3D printer.

Digital light processing [DLP] – a 3D printing process that uses liquid material, called resin. DLP 3D printers heat the resin and print it in similar fashion to an FDM 3D printer. As they work with material called resin, DLP 3D printers are also referred to as resin 3D printers.

Resin [R] – liquid material used by a DLP 3D printer.

3D bioprinting (bioprinting) [3DB] – the use of 3D printing technologies, combined with special filaments such as bioink and other biomaterials, to replicate parts that imitate bones, natural tissues, tendons and ligaments, skin, blood vessels, and even organs.

In general, bioprinting combines the creation of a scaffold (polymeric biomaterial that provides structural support for cell attachment and subsequent tissue development) with the depositing (3D printing) of living cells within the scaffold. The bioprinted tissue is then allowed to mature.

3D modeling software [3DMS] – the software for creating and editing a 3D CAD (computer-aided design) digital model of an object you want to 3D print.

Standard Tessellation Language (Stereolithography) File [STLF] – the digital file that describes the surface geometry of your object. You use 3D modeling software to create an STL file.

Slicing software [SlS] – the software that takes an STL file (the surface geometry of an object) and slices it to create the layers needed for 3D printing. It is these layers that are successively 3D printed with either filament or resin.

G-code [GC] – the file created by the slicing software. It contains a CNC (computer numerical control) set of instructions that tells your 3D printer how to print your object layer by layer.

Build area [BA] – the maximum size of an object you can 3D print. Typical dimensions for build areas include: width of 3 to 12 inches, depth of 3 to 8 inches, and height of 5 to 10 inches. When purchasing a 3D printer, the build area is an important consideration.

3D reprinting [3DR] – uses existing items to recycle filament/resin and uses that recycled filament/resin to 3D print new versions of the items.

The Periodic Table of the Fourth Industrial Revolution

Extended Reality

XR	AR	VR
VRH	MR	MRH
Hg	A	SM
DeFi	Ft	DAO
Ne	Bmm	NLP

															NFT	SiC	CW	CE	PrK	PK	PKC	C	3DR	BA	SiS	STLF	3DB			
											ST	AV	DAT	LC	D	SCi	Dapp	I	M	BHK	BN	Bc	3DR	BA	TMAI	GC	3DMS			
3DP	FDM	F									AD	eVTOL	W30	Mv	Ro	GA	DL	UL	SL	ML	ANN	ASI	SAAI							
IoT	DLP	R																												
IoCT	IoET	SH	IIoT																											
SC	SG																													
AI	ANi		RAI																											
		IoAT	LMAI	AGI	TMAI																									

														DT	HFT	FC	Ti	EC	Li	DLT	5G	N	LiFi	Q	NFC	TA			
CT	Bt													DST	ET	FC	SLAM	EC	OCR	DLT	Rdr	N	L	QC	EMV	RFID	SS		
SsT	HT													MC	B	HFC	GP	Bm	360C	WF	PvS	S	CV	CS	SR	CIC			
E	GhG													PB	Pe	OP	Hp		MoSo	WP	CSP	RE	C2	GW					
BTU	GE													SB	Ba					SP		CN							

QRC	HaT	DST	DT	FC	EC	DLT	N	Q	TA
UPC	Brc	ET	HFT	Ti	Li	5G	LiFi	NFC	RFID
PA	GR	MC	B	SLAM	OCR	Rdr	L	CV	SR
EVB	AB	PB	Pe	HFC	GP	MoSo	PvS	CS	C2
SSB	LIB	SB	Ba	OP	Bm	Hp	CSP	EH	CN
						WP	SP		GW

62

CHAPTER 10
EXTENDED REALITY

Extended reality [XR] – an umbrella term for augmented, mixed, and virtual reality, which are all some combination of the physical and virtual worlds through technology.

Augmented reality [AR] – the enhancing of the view of the physical world by adding content, information, and the like.

Augmented reality is widely used while viewing television sports programming. In swimming for example, you see different colors of lines across the pool marking world record and Olympic record time splits. Augmented reality is widely used for (American) football, showing the distance to a first down or where the team needs to get to kick a field goal.

Augmented reality is popular as well for games like Pokémon Go and other personal applications like Google Translate and even viewing the nutritional content of restaurant menu items.

To use most augmented reality applications, you need an AR-enabled viewing device such as a smartphone, tablet, computer, or television.

Virtual reality [VR] – a completely immersive simulated virtual experience, which may be based in a real world environment (e.g., walking the Great Wall of China or swimming the Great Barrier Reef off the coast of Queensland, Australia) or a fictitious one such experiencing the wreckage of Jurassic Park or leaping, sliding, running, and jumping across city rooftops (Stride).

The focus of virtual reality is to transport you via your senses to another place and set of experiences. Your senses include what you see, what you hear, what you feel, and even what you smell. (OVR Technology offers a suite of simulated smell-producing technologies.)

Virtual reality is popularly used in personal games but also includes commercial applications in engineering and manufacturing, construction, design, medicine, education, and many others.

Virtual reality headset [VRH] – a headset that covers your complete field of vision. As you move or turn in different directions, the view within the headset adjusts accordingly. VR headsets also incorporate sound technologies, giving you the ability to provide verbal instructions and also hear the sounds of the simulated environment.

Mixed reality [MR] – the merging of the physical and virtual worlds in which both worlds simultaneously co-exist in real time and are constrained by each other.

In mixed reality, the real world is mapped via some technology such as Lidar. Then, what you can do in the virtual aspect of mixed reality is constrained by the mapping of the physical world. A simple application of mixed reality is that of designing the interior of the home. The physical mapping would include the dimensions of the rooms, the location of doors and windows, etc. You could then virtually place furniture in the room, with the physical constraints not allowing you to place a piece of furniture in the room that is too big.

Mixed reality headset [MRH] – similar to virtual reality headsets, these incorporate mapping capabilities to create the appropriate interplay among the physical and virtual worlds.

Hologram [Hg] – a 3D image with depth that has been created using light beams, millions of them. Real holograms are created from a recording of a light field, not from a traditional camera lens. Using panels of lights sitting opposite of each other, two light beams from each panel intersect in space and create a pixel, a tiny dot suspended in space that you can see. You can change the color of the light beams emanating from each panel to create a different color at the intersection. Do this with enough pairs of millions of light beams and you get a real 3D hologram with depth.

Avatar [A] – a personalized graphical depiction that represents a computer user, which can be an alter ego or character that the computer user wants to portray. An avatar can be 3-dimensional (pseudo or real hologram) or 2-dimensional, as is often used in online gaming and virtual worlds. Avatars are also popular in Internet chat, messaging systems, blogs, and are very common in virtual reality.

Smart mirror [SM] – a mirror that displays your image and also extended reality content like augmented information, avatars, changes to your hair style or clothing, the correct posture for exercising, and so on.

The Periodic Table of the Fourth Industrial Revolution

Super Technologies

3DP	FDM	F						ST	AV	DAT	LC	D
IoT	DLP	R	3DB					AD	eVTOL	W30	Mv	Ro
IoCT	IoET	SH	3DMS	StS	STLF	C	PKC	PK	PrK	CE	CW	SiC	NFT	XR	AR	VR
SC	SG	IIoT	IoAT	BA	GC	3DR	Bc	BN	BHK	M	I	Dapp	SCi	VRH	MR	MRH
AI	ANI	RAI	LMAI	TMAI	AGI	SAAI	ASI	ANN	ML	SL	UL	DL	GA	Hg	A	SM
														DeFi	Ft	DAO
														Ne	Bmm	NLP

CT	BI														
SsT	HT	RFID	NFC	Q	QC	EMV	N	DLT	CC	EC	FC	DT	DST	HaT	QRC
E	GhG	SS	SR	TA		CV	WiFi	5G	6G	Li	Ti	HFT	ET	Brc	UPC
BTU	GE	CIC	CZ			CS	S	Rdr	360C	OCR	SLAM	B	MC	GR	PA
		GW	CN			EH	RE	MoSo	WF	GP	HFC	Pe	PB	AB	EVB
							SP	WP	Hp	Bm	OP	Ba	SB	LiB	SSB
							CSP								
							PvS								

66

CHAPTER 11
SUPER TECHNOLOGIES

Super technology [ST] – a technology that results as a combination of several other fourth industrial revolution technologies. They are not "super" in the sense that they are better; they are simply super in the sense that they are a combination of other technologies, and indeed they exist only as a result of a combination of several other fourth industrial revolution technologies.

Super technologies in the fourth industrial revolution include:
- Autonomous vehicle
- Drone
- Web 3.0
- Metaverse
- Robotics

Autonomous vehicle [AV] – a vehicle that can guide itself without human conduction, using sensing technologies, software (especially artificial intelligence), and a variety of driver-assistance technologies to navigate the vehicle.

Like many of the fourth industrial revolution technologies, autonomous vehicles are arriving in stages or levels. With respect to autonomous vehicles, the Society of Automotive Engineers has defined the following levels.

0 - No automation at all, although warnings for things like lane departure may be present.
1 – Some support for steering (e.g., lane centering) <u>or</u> braking and accelerating.
2 – Some support for steering <u>and</u> braking and accelerating.
3 – The car can drive itself under certain limited conditions. The car may require that you take control of driving.
4 – Same as #3 (driving under limited conditions), except that the car will not require that you take control of driving.
5 – The car can drive itself everywhere and under all conditions.

As a super technology, an autonomous vehicle is best depicted as a formulaic combination of other fourth industrial revolution technologies.

$$AV = AI + SsT + CT + IoT + E$$

Where:

- Artificial intelligence (AI): Pre-loaded AI software that has perhaps driven millions of miles in a simulated environment. This AI will also exhibit machine learning, the ability to continually learn.
- Sensing technologies (SsT): Including (1) seeing technologies such as Lidar, sonar, radar, and cameras with AI-based image recognition, (2) hearing technologies for detecting emergency vehicle sirens and the like, and (3) feeling technologies for, as examples, wind and road conditions.
- Communications technologies (CT): Intra AV within the vehicle itself for communication among the various sensors and instructions from the driver and inter AV communications technologies for communicating with other autonomous vehicles.

These may include WiFi, 5G (or perhaps 6G), and Bluetooth.

- Internet of Things (IoT): The connections among all the technologies to collect, process, and exchange data.
- Energy (E): Mostly in the form of lithium-ion batteries and solid-state batteries in the future.

As such, there are really very few – if any – technologies of the fourth industrial revolution specific to autonomous vehicles.

However, you do see 2 terms commonly associated with autonomous vehicles – driver-assistance technologies and limited conditions.

Driver-assistance technologies [DAT] – control a specific aspect of the operation of an autonomous vehicle, mostly in steering, braking, and accelerating. Technically, warnings for things like lane departure and driver fatigue are not driver-assistance technologies. They warn the driver but do not take corrective action. You will, however, see these types of warnings listed under driver-assistance technologies. (The list below includes them.)

- Lane departure warning - usually a beeping sound when the vehicle notices the driver is "drifting." Some vehicles also provide haptic feedback by making the steering wheel vibrate.
- Driver fatigue warning – usually a beeping sound when the vehicle notices that the driver seems to be steering in a less-than-optimal fashion (e.g., weaving within a lane).
- Blind spot warning – usually a beeping sound when the driver is steering in the direction of an object or an object is moving toward the vehicle.

- Automatic braking (collision avoidance) – when objects suddenly appear in front of or behind the vehicle.
- Adaptive cruise control – the driver sets the speed and the distance to stay behind vehicles in front. The vehicle will slow down when it gets within a certain distance of the vehicle in front of it.
- Parallel parking – no explanation necessary.
- Weather-related assistance – controlling wind shield wipers based on moisture, moving between dim and bright lights based on fog/rain/snow, shifting into or out of all-wheel or 4-wheel drive based on snow and ice.
- Remote parking – without the driver in the vehicle, it will navigate getting into and out of tight parking spots.
- Hands-free navigation/steering – no explanation necessary.

Limited conditions [LC] – the conditions under which an autonomous vehicle will conduct itself. This is definitely important within the context of the arriving stages of autonomous vehicles. Limited conditions can include:
- Highway driving only (i.e., not on busy city streets)
- (Good) weather
- (Minimal or no) traffic congestion
- (No) construction areas in which speed is limited.

They make sense. A long stretch of highway in good weather with little other traffic and no construction is optimal (right now) for an autonomous vehicle to fully take control of steering, accelerating, braking, shifting lanes, etc.

Drone [D] – flying vehicle without needing a pilot in the vehicle. A drone may be remote-controlled or a completely autonomous drone.

Autonomous drone [AD] – aircraft with the necessary sensing technologies and artificial intelligence to fly itself without human conduction.

As a super technology and similar to an autonomous vehicle, a drone is best depicted as a formulaic combination of other fourth industrial revolution technologies.

$$AV = AI + SsT + CT + IoT + E$$

Where:
- Artificial intelligence (AI): May or may not be present in a drone today, but certainly will be in the future. For a drone to be autonomous, it must have sophisticated AI, just as an autonomous vehicle must.
- Sensing technologies (SsT): Including (1) seeing technologies such as Lidar, sonar, radar, and cameras with AI-based image recognition, (2) feeling technologies for, as examples, wind and wind shear, both horizontally and vertically, and perhaps (3) smelling technologies for detecting the presence of smoke and fire.
- Communications technologies (CT): Even the most basic drones require communications, at a minimum, from the remote control you operate to the drone itself. And eventually, we'll need drone-to-drone communications, much like autonomous vehicles will communicate with each other to make collective decisions.

- Internet of Things (IoT): The connections among all the technologies to collect, process, and exchange data.
- Energy (E): Mostly in the form of lithium-ion batteries and solid-state batteries in the future.

As such, there are really very few – if any – technologies of the fourth industrial revolution specific to drones. But, you do see a new term here, eVTOL.

Electric vertical take-off and landing [eVTOL] – refers to the ability of an electric-powered aircraft to take off, hover, and land vertically. So, eVTOL aircraft do include personal and (most) commercial drones, and can also include air taxis, basically the equivalent of a flying car.

Flying cars hold a lot of promise… reduced traffic congestion on the roads, must faster "as the crow flies" times from suburban housing to city-center workplaces (and back), reduction in carbon emissions because of the electric power, faster response times to accidents, etc.

Flying car initiatives include City Airbus NexGen (by Airbus), SkyDrive (a Japanese startup backed by Toyota), Joby Aviation, Hyundai and Supernal flying taxi, Boeing, and AirCar.

Web 3.0 [W30] – the next evolution of the Web beyond its current form that will exhibit semantic characteristics, decentralized protocols, and ubiquity.

As a super technology, the Web 3.0 is best depicted as a formulaic combination of other fourth industrial revolution technologies.

$$W30 = AI + DLT + IoAT$$

- Artificial intelligence (AI) and semantic characteristics: Semantic refers to a concept called metadata, essentially data about data. Using metadata, Web 3.0 applications will be able to – in automated fashion – connect related data sources from all over the world. This will require the use of artificial intelligence to connect, make sense of, and interpret the data. For example, consider I love Bitcoin, I <3 Bitcoin, and I heart Bitcoin. Syntax and use of words and symbols are different; the semantic meaning of them is essentially the same. We can understand that; computers will need AI to do so.
- Distributed ledger technology (DLT) and decentralized protocols: The current Web is heavily dominated by the centralization of storage, processing, and control of information. Web 3.0 will decentralize information and software using distributed ledger technologies like blockchain, cryptocurrency, tokens such as NFTs, smart contracts and Dapps that automate tasks, and decentralized financial (DeFi) applications.
- The Internet of All Things (IoAT) and ubiquity: Ubiquity simply means everywhere. For the Web 3.0 to be everywhere, everything must be connected to the Web. Which means that the Internet of <u>All</u> Things will play an important role in the foundation of the

Web 3.0. Connecting all things – computer-related, electronic, and non-electronic – to the Web will create a Web 3.0 that is everywhere.

We're definitely not at Web 3.0 yet. As we evolve to Web 3.0, you probably won't even notice the changes as they occur. They will be subtle and small. But, over the next 10 years, the Web will change dramatically, with all those subtle and small changes accumulating into a significant shift in the Web.

Metaverse [Mv] – a virtual universe – aided by technologies such as augmented/mixed/virtual reality, holograms and avatars, various sensing technologies, and decentralized protocols – in which people connect, live, work, conduct commerce, and play.

As a super technology, metaverse is best depicted as a formulaic combination of other fourth industrial revolution technologies.

$$MV = XR + Hg + A + SsT + DLT$$

- Extended reality (XR): Including augmented, mixed, and virtual reality that will transport the user to another (virtual) place.
- Holograms (Hg) and avatars (A): Simulations of people and objects in the metaverse.
- Sensing technologies(SsT): Various technologies that will (1) capture, interpret, and respond to the movements of the user and (2) create sight, sound, haptic feedback, and perhaps even smell to engage the user in a more immersive experience.
- Distributed ledger technologies (DLT) and decentralized protocols: In most implementations of the metaverse concept, people need money to buy land, purchase cars, play games, and conduct

commerce. These will require the use of DLT concepts like blockchain, cryptocurrency, NFTs, and decentralized finance (DeFi) applications.

The goal of the metaverse concept is to connect people through a more immersive experience than traditional technology tools like social media and team collaboration tools.

Robotics [Ro] – a branch of engineering that focuses on the design, construction, and use of robots (machines) to assist humans in performing tasks.

The field of robotics is vast, with a variety of implementations. You can think of a 3D printer as a simple robot. It uses g-code instructions to render a 3D object. Similar types of robots/machines are found throughout manufacturing environments, expertly executing a set of instructions 24x7.

There are also advanced robotic applications and machines that utilize artificial intelligence and image/sound/smell recognition to respond to commands, identify objects, and perform a variety of non-repetitive tasks.

FINAL THOUGHTS

I do love the periodic table approach to thinking about and visualizing big topics, such as the fourth industrial revolution. For me, it's the opportunity to ponder organization, relationships, and dependencies.

And, it's been a wild ride. Organizing, creating abbreviations, determining what to include and what to leave out, and laying out the periodic table have all been real challenges.

This is the first I've seen for the fourth industrial revolution. So, my work is at least new, and hopefully breaking some new ground.

Just as the periodic table of elements has changed over time, I'm sure the same will be true for the periodic table of the fourth industrial revolution.

Who knows, perhaps metaverse will wither and fade away. And there will most certainly be new technologies and terms that emerge in the fourth industrial revolution.

I would love to get your feedback. Please feel free to contact me at techbookwriter@gmail.com if you have any suggestions, additions, revisions, and/or comments.

INDEX

360 camera [360C], 24
3D bioprinting [3DB], 60
3D modeling software [3DMS], 60
3D printing [3DP], 59
3D reprinting [3DR], 61
5G [5G], 18
6G [6G], 18

A
Alkaline battery [AB], 12
Artificial general intelligence [AGI], 43
Artificial intelligence [AI], 41
Artificial narrow intelligence [ANI], 42
Artificial neural network [ANN], 44
Artificial superintelligence[ASI], 44
Augmented reality [AR], 63
Autonomous drone [AD], 71
Autonomous vehicle [AV], 67
Avatar [A], 65

B
Barcode [Brc], 26
Battery [Ba], 11
Biomass [Bm], 10
Biometrics [B], 25
Biomimicry [Bmm], 47
Block hash or key [BHK], 50
Blockchain [Bc], 49
Blockchain node [BN], 49
Bluetooth [Bt], 15
British thermal unit [BTU], 6
Build area [BA], 61

C
Carbon neutral [CN], 7
Carbon sequestration [CS], 7
Carbon zero [CZ], 7
Climate change [ClC], 7
Cloud computing [CC], 35
Communications technology [CT], 15
Computer vision [CV], 23
Concentrated solar power [CSP], 9
Cryptocurrency [C], 55
Cryptocurrency exchange [CE], 56
Cryptocurrency wallet [CW], 56

D
Decentralized app [Dapp], 50
Decentralized finance [DeFi], 51
Decentralized autonomous organization [DAO], 51
Deep learning [DL], 46
Digital light processing [DLP], 59
Digital scent technology [DST], 28
Digital twin [DT], 28
Distributed ledger technology [DLT], 33
Driver-assistance technology [DAT], 69
Drone [D], 71

E

Edge computing [EC], 35
Electric vertical take-off and landing [eVTOL], 72
Electric-vehicle battery [EVB], 12
Energy [E], 5
Energy harvesting [EH], 8
EuroPay-Mastercard-Visa [EMV], 17
Extended reality [XR], 63
Eye tracking [ET], 26

F

Filament [F], 59
Fintech [Ft], 51
Fog computing [FC], 35
Fused deposition modeling [FDM], 59

G

G-code [GC], 60
Genetic algorithm [GA], 47
Geothermal power [GP], 10
Gesture recognition [GR], 26
Global warming [GW], 7
Greenhouse effect [GE], 7
Greenhouse gas [GhG], 6

H

Hand and finger tracking [HFT], 26
Haptic technology [HaT], 27
Hearing technology [HT], 22
Hologram [Hg], 65
Hydrogen fuel cell [HFC], 10
Hydropower [Hp], 9

I

Immutability [I], 50

Internet of All Things [IoAT], 39
Industrial Internet of Things [IIoT], 38
Internet of Computer Things [IoCT], 37
Internet of Electronic Things [IoET], 37
Internet of Things [IoT], 37

L

Latency [Lt], 18
Lidar [L], 23
Light fidelity [LiFi], 17
Limited conditions [LC], 70
Limited-memory artificial intelligence [LMAI], 43
Lithium-ion battery [LIB], 12

M

Machine learning [ML], 45
Mining [M], 50
Metaverse [Mv], 74
Mixed reality [MR], 64
Mixed reality headset [MRH], 65
Mobile solar [MoSo], 9
Motion capture [MC], 25

N

Nano [N], 33
Nanotechnology [Nt], 33
Natural language processing [NLP], 47
Near-field communication [NFC], 16
Neuroevolution [Ne], 47
Non-fungible token [NFT], 57

O
Ocean power [OP], 10
Optical character recognition [OCR], 24

P
Paris Agreement [PA], 11
Photovoltaic system [PvS], 8
Piezoelectricity [Pe], 10
Primary battery [PB], 12
Private key [PrK], 55
Public key [PK], 55
Public key cryptography [PKC], 55

Q
Quantum computing [QC], 32
Qubit [Q], 32
Quick response code [QRC], 27

R
Radar [Rdr], 23
Radio frequency identification [RFID], 16
Reactive artificial intelligence [RAI], 42
Renewable energy [RE], 6
Resin [R], 60
Robotics [Ro], 75

S
Secondary battery [SB], 12
Self-aware artificial intelligence [SAAI], 44
Sensing technology [SsT], 21
Simultaneous localization and mapping [SLAM], 25
Slicing software [SlS], 60
Smart city [SC], 38
Smart contract [SCt], 50
Smart grid [SG], 38
Smart home [SH], 38
Smart mirror [SM], 65
Solar power [SP], 8
Solid-state battery [SSB], 13
Sonar [S], 23
Sound sensor [SS], 22
Speech recognition [SR], 22
Stablecoin [StC], 57
Standard Tessellation Language file [STLF], 60
Super technology [ST], 67
Supervised learning [SL], 46

T
Technology architecture [TA], 31
Tethering [Tt], 18
Theory of mind artificial intelligence [TMAI], 43

U
Universal product code [UPC], 26
Unsupervised learning [UL], 46

V
Virtual reality [VR], 64
Virtual reality headset [VRH], 64

W
Web 3.0 [W30], 73
Wind farm [WF], 9
Wind power [WP], 9
Wireless fidelity [WiFi], 17

ABOUT THE AUTHOR

Stephen Haag is a Professor of the Practice in the Daniels College of Business Department of Business Information & Analytics at the University of Denver. Stephen received his MBA from West Texas State University in 1998 and his PhD from the University of Texas at Arlington in 1992. After joining the University of Denver in 1995, Stephen has held the positions of Chair of the Department of Information Technology and Electronic Commerce, the Director of Daniels Technology, the Director of the MBA, Associate Dean of Graduate Programs, the Director of Assurance of Learning, and the Director of Entrepreneurship.

Stephen is the author or co-author of 49 books, including *The Fourth Industrial Revolution 2022: What Every College and High School Student Needs to Know About the Future*.

He lives in Highlands Ranch, Colorado with his wife Pam, four children (Darian, Trevor, Katrina, and Alexis) and 2 rescue shelter dogs (Loki and Cricket).

Made in the USA
Columbia, SC
11 July 2022